The Beginner's Guide to Raising Chickens

Everything You Need to Know to Raise a Thriving Flock

MACY HUM

Table of Contents

Introduction

Have you ever dreamed of owning your own chickens either in the farm or backyard? Do you want to enjoy fresh eggs, organic fertilizer, and feathered friends in your backyard? Do you want to learn more about these amazing animals and their fascinating history, evolution, and behavior? This book is for you if any of these questions apply to you. The Beginner's Guide to Raising Chickens is a comprehensive and practical guide that will teach you everything you need to know to start and succeed in your chicken-keeping adventure. Whether you are a complete beginner or a seasoned pro, you will find useful tips, tricks, and insights in this book.

In this book, you will discover:
- How to choose the right breed, number, and age of chickens for your needs, space, and budget.
- How to set up a cozy and secure coop and run for your chickens, with all the essential features and accessories.

- How to care for your chickens from day one to adulthood, including feeding, watering, cleaning, protecting, and treating them.
- How to collect and store your eggs, and how to deal with common problems and solutions related to your eggs.
- How to interact with your chickens and learn their personalities, and how to train and entertain them with toys and treats.
- How to use your chickens' manure as a valuable resource for your garden, compost, or biogas.
- How to appreciate the amazing features and functions of your chickens' body, and how to understand their behavior and communication.
- How to trace the origins and development of your chickens, and how to explore their genetic and geographic diversity.
- How to appreciate the cultural and religious significance of chickens around the world, and how to join a chicken community and share your experiences.

By the end of this book, you will have all the knowledge and skills you need to raise a happy and healthy flock of chickens. You will also have a deeper appreciation and love for these wonderful creatures, and the rewards they offer.

So what are you waiting for? Grab this book today, and get ready to embark on your chicken-keeping journey. Your chickens are waiting for you!

The Goals and Benefits of Raising Chickens

To have a source of organic eggs and meat

One of the main reasons why people raise chickens is to have a source of organic eggs and meat. Eggs and meat from backyard chickens are fresher, tastier, and more nutritious than those from factory farms. They are also free from hormones, antibiotics, and chemicals that may harm your health and the environment. By raising your own chickens, you can control what they eat, how they live, and how they are processed. You can also choose the breeds that suit your preferences and needs. For example, some breeds are more productive, some are more colorful, and some are more friendly.

You can enjoy the satisfaction and pride of producing your own food, and share it with your family and friends.

To earn some extra cash from selling eggs, meat, or manure

Another benefit of raising chickens is that you can earn some extra cash from selling eggs, meat, or manure. Depending on the demand and supply in your area, you can sell your eggs and meat at a higher price than the market, especially if they are organic and free-range. You can also sell your chicken manure, which is a valuable fertilizer, compost, or biogas. You can market your products online, through word-of-mouth, or at local farmers' markets, fairs, or shops. You can also barter your products for other goods or services. By selling your products, you can recover some of the costs of raising chickens, and even make a profit.

To reduce food waste by feeding chickens kitchen scraps

A third benefit of raising chickens is that you can reduce food waste by feeding chickens kitchen scraps. Chickens are omnivorous animals, which means they can eat both plant and animal matter. They can eat most of the leftovers, peelings, and trimmings from your kitchen, as well as fruits, vegetables, grains, seeds, nuts, and insects from your garden. They can also eat some dairy products, such as cheese, yogurt, and milk, and some meat products, such as fish, liver, and bones. However, you should avoid feeding them anything that is spoiled, moldy, salty, spicy, or toxic, such as chocolate, onion, garlic, avocado, or citrus. By feeding your chickens kitchen scraps, you can save money on feed, reduce landfill waste, and provide a varied and balanced diet for your chickens.

To improve soil quality and fertility by using chicken manure as fertilizer, compost, or biogas

A fourth benefit of raising chickens is that you can improve soil quality and fertility by using chicken manure as fertilizer, compost, or biogas.

Fertilized chicken dung is an excellent source of nitrogen, phosphorus, potassium, and other elements that plants require for healthy growth. It also contains organic matter that improves soil structure, water retention, and microbial activity. You can use chicken manure as a fertilizer, by spreading it directly on your garden beds, or as a compost, by mixing it with other organic materials, such as leaves, grass clippings, or straw, and letting it decompose. You can also use chicken manure as a biogas, by converting it into methane gas, which can be used for cooking, heating, or lighting. By using chicken manure, you can enhance your garden's productivity, diversity, and beauty.

To control pests and weeds by letting chickens forage and scratch

A fifth benefit of raising chickens is that you can control pests and weeds by letting chickens forage and scratch. Chickens are natural pest controllers, as they love to eat insects, worms, snails, slugs, and other pests that may damage your crops or plants. They also help to aerate and loosen the soil, by scratching and digging with their feet and beaks. This helps to expose and eliminate weed seeds, roots, and rhizomes, and to incorporate organic matter into the soil.

By letting your chickens forage and scratch, you can reduce your reliance on pesticides and herbicides, and improve your soil health and fertility.

To enhance biodiversity and ecology by providing habitat and food for wildlife

A sixth benefit of raising chickens is that you can enhance biodiversity and ecology by providing habitat and food for wildlife. Chickens are part of the natural ecosystem, and they can coexist and interact with other animals and plants. By raising chickens, you can provide shelter, food, and water for various birds, mammals, reptiles, amphibians, and insects that may visit or live in your backyard. For example, you can attract songbirds, hummingbirds, butterflies, bees, and other pollinators, by planting flowers, herbs, and shrubs around your coop and run. You can also attract owls, hawks, foxes, raccoons, and other predators, by leaving some eggs, meat, or bones for them. However, you should also protect your chickens from these predators, by using fences, nets, or electric wires. By providing habitat and food for wildlife, you can increase the diversity and balance of your backyard ecosystem.

To learn more about chicken anatomy, physiology, behavior, and intelligence

A seventh benefit of raising chickens is that you can learn more about chicken anatomy, physiology, behavior, and intelligence. Chickens are fascinating animals, and they have many features and functions that you may not be aware of. For example, did you know that chickens have a four-chambered heart, a complex digestive system, and a sophisticated immune system? Did you know that chickens have a pecking order, a social hierarchy, and a communication system? Did you know that chickens can recognize faces, solve problems, and remember events? By raising chickens, you can observe and study these aspects of chicken biology and psychology, and gain a deeper understanding and appreciation of these amazing creatures.

To explore chicken history, evolution, and diversity

An eighth benefit of raising chickens is that you can explore chicken history, evolution, and diversity. Chickens have a long and rich history, and they have evolved and diversified over thousands of years. Chickens are believed to be descended from the red junglefowl, a wild bird that lives in Southeast Asia. They were domesticated by humans around 8,000 years ago, and they spread across the world through trade, migration, and colonization. Today, there are more than 300 breeds of chickens, each with its own characteristics, origins, and purposes. Some breeds are more ancient, some are more modern, and some are more rare. By raising chickens, you can discover and appreciate the history, evolution, and diversity of these remarkable birds.

To appreciate chicken culture and religion around the world

A ninth benefit of raising chickens is that you can appreciate chicken culture and religion around the world.

Chickens are not only animals, but also symbols and icons of various cultures and religions. Chickens have been used in rituals, ceremonies, and festivals, to celebrate, worship, or honor different gods, spirits, or ancestors. Chickens have also been associated with different meanings, values, or attributes, such as fertility, prosperity, courage, or wisdom. Chickens have also inspired art, literature, music, and folklore, in different forms and styles. By raising chickens, you can explore and appreciate the cultural and religious significance of chickens around the world, and how they reflect the diversity and creativity of human civilization.

To bond with chickens and enjoy their personalities and companionship

A tenth and final benefit of raising chickens is that you can bond with chickens and enjoy their personalities and companionship. Chickens are not just food, but also friends. Chickens have distinct personalities, and they can show emotions, such as happiness, sadness, anger, or fear. Chickens can also form bonds with humans, and they can recognize, trust, and love their owners. Chickens can also provide entertainment, relaxation, and therapy, for their owners.

By raising chickens, you can develop a close and meaningful relationship with your chickens, and enjoy their company and affection.

The Challenges and Responsibilities of Keeping Backyard Chickens

Noise

Chickens can make loud noises, especially roosters, which can crow at any time of the day or night. This can annoy your neighbors or violate noise ordinances in your area. Noise can also attract unwanted attention from predators or thieves. To reduce noise, you can choose quieter breeds, such as bantams, or avoid keeping roosters altogether. You can also soundproof your coop and run, by using insulation, curtains, or plants. You can also talk to your neighbors and explain the benefits of keeping chickens, and maybe offer them some eggs or manure as a gesture of goodwill.

Odor

Chicken poop has a strong and unpleasant smell, which can be offensive to you or your neighbors. Odor can also attract flies, maggots, or rodents, which can spread diseases or parasites. To reduce odor, you can clean your coop and run regularly, and remove any wet or dirty bedding. You can also use absorbent and deodorizing materials, such as wood shavings, straw, or sand. You can also compost your chicken manure, by mixing it with other organic materials, such as leaves, grass clippings, or straw, and letting it decompose. This will reduce the smell and volume of the manure, and turn it into a valuable fertilizer.

Predators

Chickens are vulnerable to attacks from various animals, such as foxes, raccoons, hawks, or dogs, which can harm or kill them. Predators can also damage or destroy your coop and run, and steal your eggs or feed. To protect your chickens from predators, you can use fences, nets, or electric wires, to enclose your coop and run. You can also use locks, latches, or hinges, to secure your doors and windows. You can also use lights, alarms, or motion sensors, to deter or scare away predators.

You can also keep your chickens indoors at night, or provide them with hiding places, such as bushes, boxes, or tunnels.

Time and commitment

Chickens need daily care, such as feeding, watering, and egg collection, as well as regular cleaning and maintenance of their coop and run. They also need occasional care, such as trimming their nails, dusting their feathers, or checking their health. Chickens also need attention and affection, as they are social and curious animals. To care for your chickens, you need to dedicate some time and commitment, and plan ahead for any contingencies. You need to have a reliable and consistent routine, and follow the best practices and tips for chicken-keeping. You also need to find a trustworthy and experienced chicken sitter, who can take care of your chickens when you are away or busy.

Zoning restrictions

Some areas may have laws or regulations that limit or prohibit keeping backyard chickens, such as the number, type, or location of chickens. These laws or regulations may vary depending on your city, county, or state, and they may change over time.

To comply with zoning restrictions, you need to do some research and check the current and updated rules and requirements for your area. You may need to obtain a permit, license, or registration, to keep backyard chickens. You may also need to follow certain standards or guidelines, such as the size, design, or placement of your coop and run. You may also need to respect the rights and interests of your neighbors and community.

The Basic Requirements and Expectations of Chicken-Keeping

- *Planning:* Before starting to raise chickens, you need to consider the type, breed, and size of the chickens you want, the legal and ethical issues involved, and the goals and benefits you expect from chicken-keeping. You also need to research the best practices and methods for your specific situation and location.

- *Resources:* Chickens need a secure and comfortable coop, a spacious and fenced run, clean and fresh water, nutritious and balanced feed, and other supplies such as bedding, nest boxes, roosts, feeders, waterers, and grit. You need to obtain or prepare these resources before acquiring the chickens, and make sure they are adequate and appropriate for your flock.

- *Maintenance:* Chickens need daily, weekly, and monthly care to ensure their health, happiness, and productivity. You need to provide them with fresh water and feed, collect their eggs, observe their behavior and appearance, and clean their coop and run. You also need to prevent and treat common problems such as diseases, parasites, predators, and pests.

- *Welfare:* Chickens are sentient animals that deserve respect and compassion. You need to follow the code of accepted farming practice for the welfare of poultry, which includes providing them with adequate space, light, ventilation, shelter, food, water, and social contact. You also need to avoid causing them unnecessary pain, stress, or suffering.

These are the basic requirements and expectations of chicken-keeping that you need to fulfill if you want to enjoy the rewards and benefits of this hobby.

Part I: Getting Ready for Your Chickens

Selecting the Right Breed, Number, and Age

If you are thinking of raising backyard chickens, you might be wondering how to choose the right breed, number, and age of your flock. There are many factors to consider, such as your climate, space, budget, egg production, personality, and appearance preferences. Here are some tips to help you make the best decision for your situation.

Breed

There are hundreds of chicken breeds, each with their own characteristics and traits. Some breeds are more cold-hardy, heat-tolerant, friendly, docile, noisy, or colorful than others. Some breeds lay more eggs, larger eggs, or different colored eggs than others. Some breeds are more suitable for meat production, ornamental purposes, or dual-purpose (both eggs and meat). To select the right breed for you, you need to ask yourself what are your main goals and expectations from your chickens. Do you want them to provide you with fresh eggs, meat, or both? Do you want them to be friendly and sociable, or more independent and aloof?

Do you want them to be quiet and discreet, or loud and proud? Do you want them to have a certain look or color?

You can research different breeds online or by visiting local farms or poultry shows. You can also ask other chicken owners for their recommendations and experiences. Here we are going to mention Some popular breeds for your backyard chickens. They are as follows:

Rhode Island Red
A hardy, productive, and friendly breed that lays large brown eggs. They are adaptable to different climates and environments, and have a reddish-brown plumage.

Australorp
A calm, gentle, and docile breed that lays medium to large brown eggs. They are good foragers and can tolerate both hot and cold weather. They have a black plumage with a greenish sheen.

Buff Orpington
A fluffy, cuddly, and affectionate breed that lays medium to large brown eggs. They are good mothers and can go broody (sit on their eggs to hatch them). They are cold-hardy and have a buff-colored plumage.

Leghorn

A lively, active, and flighty breed that lays large white eggs. They are excellent layers and can produce up to 300 eggs per year. They are heat-tolerant and have a white plumage with a red comb and wattles.

Easter Egger

A fun, colorful, and friendly breed that lays medium to large eggs in various shades of blue, green, pink, or brown. They are not a true breed, but a mix of different breeds that carry the blue egg gene. They have a diverse plumage that can range from black, white, brown, red, or gray.

Number of Chickens to Raise

The number of chickens you need depends on how many eggs you want, how much space you have, and how much time and money you can spend on them. Generally, the more chickens you have, the more eggs you get, but also the more space, feed, water, bedding, and care they require.

A good rule of thumb is to have at least two chickens, as they are social animals and need company. However, you should not have more than 10 chickens per person, as that might be too many to handle and consume.

You should also check your local laws and regulations regarding the maximum number of chickens you can keep in your area. Some factors to consider when deciding how many chickens to get are:

- *Egg consumption:* How many eggs do you and your family eat per week? How many eggs do you want to sell, give away, or store? On average, a chicken lays about four eggs per week, but this can vary depending on the breed, age, season, and health of the chicken.

- *Space:* How much space do you have for your chickens? How big is your coop and run? How much free range area do you have? Chickens need enough space to roam, forage, dust bathe, and exercise. A general guideline is to provide at least 4 square feet of coop space and 10 square feet of run space per chicken. However, more space is always better, as it reduces stress, boredom, and aggression among the flock.

- *Budget:* How much money can you afford to spend on your chickens? How much does the feed, water, bedding, coop, run, fencing, and other supplies cost?

How much does the vet, medication, and vaccination cost? How much does the electricity, heating, and lighting cost? Chickens can be expensive to raise, especially if you want to provide them with high-quality and organic products. You should make a budget and plan ahead before you get your chickens.

- *Time:* How much time can you devote to your chickens? How much time does it take to feed, water, clean, collect eggs, and check on them daily? How much time does it take to maintain, repair, and upgrade the coop and run? How much time does it take to interact, play, and train your chickens? Chickens can be time-consuming to care for, especially if you have a large flock or if you encounter any problems or emergencies. You should make sure you have enough time and energy to look after your chickens properly.

Age

The age of the chickens you get depends on how much experience you have, how much work you want to do, and how much patience you have.

You can get chickens at different stages of their life cycle, such as chicks, pullets, or hens.

- **Chicks:** Chicks are baby chickens that are less than 8 weeks old. They are cute, fluffy, and adorable, but they also require a lot of attention, care, and equipment. You need to provide them with a brooder, a heat source, a thermometer, a feeder, a waterer, a bedding, and a chick starter feed. You need to monitor their temperature, humidity, ventilation, and health constantly. You need to clean their brooder and change their bedding regularly. You need to handle them gently and frequently to socialize them. You also need to wait for several months before they start laying eggs. The advantages of getting chicks are that you can choose the breed, sex, and color you want, you can raise them from scratch, and you can bond with them more easily. The disadvantages are that they are more fragile, vulnerable, and demanding, they can be noisy and messy, and they can be hard to sex accurately (unless you get sex-linked or sexed chicks).

- **Pullets:** Pullets are young female chickens that are between 8 and 20 weeks old. They are more mature, independent, and hardy than chicks, but they are not yet laying eggs. You need to provide them with a coop, a run, a feeder, a waterer, a bedding, and a grower feed. You need to check on them daily and clean their coop and run weekly. You need to handle them occasionally to tame them. You also need to wait for a few more weeks before they start laying eggs. The advantages of getting pullets are that you can skip the brooding stage, you can avoid the rooster risk, you can see their personality and appearance more clearly, and you can get eggs sooner. The disadvantages are that they are more expensive, less available, and less customizable than chicks, they can be harder to integrate with an existing flock, and they can be less friendly and trusting than chicks.

- *Hens:* Hens are adult female chickens that are older than 20 weeks and are already laying eggs. They are fully grown, established, and productive, but they are also more set in their ways and habits. You need to provide them with a coop, a run, a feeder, a waterer, a bedding, and a layer feed. You need to check on them daily and clean their coop and run weekly. You need to handle them rarely to respect their space. You can enjoy their eggs immediately. The advantages of getting hens are that you can skip the growing stage, you can get instant gratification, you can see their performance and quality more accurately, and you can learn from their experience and wisdom. The disadvantages are that they are more costly, scarce, and fixed than chicks or pullets, they can be harder to integrate with an existing flock, and they can be less adaptable and sociable than chicks or pullets.

I hope you are enjoying this book, leave a review to motivate this publisher please!

Pros and Cons of Chicks, Pullets, or Hens

As we have seen in the previous section, there are different pros and cons of getting chicks, pullets, or hens. Below is a summary of the main advantages and disadvantages of each option:

Feature	Chicks	Older Chickens
Selection	More options for breed, sex, and color	Limited selection based on availability
Nurturing	Requires more care and attention	Less demanding and independent
Bond	Easier to form a strong bond	Bond may take more time and effort
Cost	More affordable and customizable	Higher upfront cost
Housing	Requires a brooder for warmth and protection	No special housing requirements
Noise and Mess	Can be noisy and messy	Generally quieter and less messy
Sexing	Can be difficult to sex accurately	Sex is usually evident
Availability	More readily available	May be less available, especially for specific breeds

If you want to start or expand your backyard flock, you need to know where and how to obtain chickens. There are different sources and methods for getting chickens, each with their own advantages and disadvantages. These few of most typical ones:

Hatcheries

Hatcheries are commercial facilities that breed, incubate, and ship chickens to customers. You can order chicks, pullets, or hens from hatcheries online, by phone, or by mail. You can choose from a wide variety of breeds, colors, and sizes. You can also request sexed, vaccinated, or certified organic chickens. The hatcheries will ship the chickens to your nearest post office, where you can pick them up.

The pros of getting chickens from hatcheries are that you can get large quantities, high quality, and guaranteed sex of chickens. You can also get rare or exotic breeds that are not available locally. The cons are that you have to pay for shipping, handling, and minimum order fees. You also have to deal with the stress and risk of shipping live animals, which can cause mortality, injury, or disease.

Local Farms

Local farms are small-scale operations that raise chickens for eggs, meat, or both. You can buy chicks, pullets, or hens from local farms directly, or through farmers' markets, feed stores, or online platforms. You can see the chickens in person, inspect their health and condition, and ask questions to the farmers. Additionally, you can help the neighborhood and local economy.

The pros of getting chickens from local farms are that you can get fresh, healthy, and happy chickens. You can also get lower prices, no shipping fees, and no minimum order requirements. The cons are that you have to travel to the farms, which can be far or inconvenient. You also have to deal with the availability, selection, and quality of the chickens, which can vary depending on the season, location, and farmer.

Hatching Eggs

Hatching eggs are fertilized eggs that you can incubate and hatch yourself. You can buy hatching eggs from hatcheries, local farms, or online platforms. You can choose from different breeds, colors, and sizes of eggs. You can also request specific traits, such as sex-linked, auto-sexing, or blue egg genes.

You need to have an incubator, a candling device, and a brooder to hatch and raise the chicks.

The pros of getting chickens from hatching eggs are that you can experience the joy and wonder of hatching your own chicks. You can also get more control, involvement, and satisfaction in the process. The cons are that you have to invest in the equipment, time, and effort to hatch and care for the chicks. You also have to deal with the uncertainty, difficulty, and failure of hatching, which can result in low hatch rates, unsexed chicks, or dead embryos.

How to Setting Up Your Chicken Coop and Run

One of the most important aspects of raising backyard chickens is providing them with a safe, comfortable, and functional coop and run. A coop is the enclosed shelter where the chickens sleep, nest, and roost. A run is the fenced area where the chickens can exercise, forage, and socialize. A good coop and run will keep your chickens healthy, happy, and productive. Here are some tips on how to design, build, or buy your coop and run, what features to include, and where to locate them.

Design, Build, or Buy
There are three main options for getting your coop and run: designing and building them yourself, buying them ready-made, or using a combination of both.

Each option has its own pros and cons, depending on your budget, skills, time, and preferences.

- *Design and build:* If you are handy, creative, and adventurous, you can design and build your own coop and run from scratch. You can use wood, metal, plastic, or recycled materials, and customize them to your liking. You can also follow online tutorials, plans, or books for inspiration and guidance. The advantages of this option are that you can save money, express your personality, and have more control and satisfaction. The disadvantages are that you have to spend more time, effort, and tools, and deal with the challenges and risks of construction.

- *Buy:* If you are busy, inexperienced, or cautious, you can buy your coop and run ready-made from online or local stores. You can choose from a variety of sizes, styles, and prices, and have them delivered and assembled for you. You can also modify or upgrade them later if you want. The advantages of these options are that you can save time, hassle, and skills, and get a professional and reliable product.

The disadvantages are that you have to spend more money, compromise your taste, and have less involvement and pride.

- **Combine:** If you are somewhere in between, you can use a combination of both options. You can buy a basic coop and run, and then add your own touches and improvements. You can also use a pre-existing structure, such as a shed, barn, or dog house, and convert it into a coop and run. The advantages of this option are that you can balance the cost, quality, and convenience, and get the best of both worlds. The disadvantages are that you still have to do some work, research, and planning, and deal with the compatibility and durability issues.

Essential Coop and Run Features

Regardless of which option you choose, your coop and run should have some essential features to ensure the well-being and productivity of your chickens.

These are a handful of the more significant ones:

- **_Space:_** Your coop and run should have enough space for your chickens to move, stretch, and explore. A general guideline is to provide at least 4 square feet of coop space and 10 square feet of run space per chicken. However, more space is always better, as it reduces stress, boredom, and aggression among the flock.

- **_Ventilation:_** Your coop and run should have adequate ventilation to allow fresh air, light, and temperature regulation. You can use windows, doors, vents, or fans to create airflow and prevent moisture, ammonia, and heat buildup. You should also make sure the ventilation is secure and predator-proof, and that it does not create drafts or cold spots.

- **_Insulation:_** Your coop and run should have sufficient insulation to keep your chickens warm in the winter and cool in the summer. You can use materials such as wood, straw, foam, or fiberglass to insulate the walls, floor, and roof of your coop.

You should also provide shade, water, and misters for your run in the hot weather, and heaters, lamps, or blankets for your coop in the cold weather.

- *Nesting boxes:* Your coop should have enough nesting boxes for your chickens to lay their eggs. A general guideline is to provide one nesting box for every four to five chickens. You can use wooden, metal, or plastic boxes, or baskets, buckets, or crates. You should also line the boxes with soft bedding, such as straw, shavings, or hay, and keep them clean and dry. You should also place the boxes in a dark, quiet, and private area of the coop, and make them accessible and comfortable for the chickens.

- *Roosts:* Your coop should have enough roosts for your chickens to perch and sleep. A general guideline is to provide at least 8 to 10 inches of roost space per chicken. You can use wooden, metal, or plastic poles, bars, or branches. You should also make sure the roosts are sturdy, smooth, and round, and that they are elevated, spaced, and angled properly.

You should also place the roosts higher than the nesting boxes, and make them easy and safe for the chickens to reach and use.

- *Feeders and waterers:* Your coop and run should have enough feeders and waterers for your chickens to eat and drink. A general guideline is to provide one feeder and one waterer for every six to eight chickens. You can use hanging, trough, or tube feeders, and nipple, cup, or bowl waterers. You should also make sure the feeders and waterers are clean, full, and fresh, and that they are protected from rain, snow, and pests. You should also place the feeders and waterers at the right height and distance for the chickens, and make them convenient and hygienic for them.

- *Litter:* Your coop and run should have enough litter for your chickens to scratch, dust bathe, and poop. You can use materials such as straw, shavings, sand, or peat moss. You should also make sure the litter is dry, deep, and loose, and that you change it regularly. You can also use the litter as a source of compost or fertilizer for your garden.

Location and Security

The location and security of your coop and run are also crucial for the safety and happiness of your chickens. You should choose a spot that is flat, level, and well-drained, and that has access to sun, shade, and wind. You should also avoid areas that are prone to flooding, erosion, or fire. You should also secure your coop and run from predators, such as dogs, cats, foxes, raccoons, hawks, or snakes. You can use materials such as wire, mesh, or electric fencing, and hardware, locks, or latches. You should also check your coop and run regularly for any gaps, holes, or weak points, and repair them as soon as possible.

Preparing for Your Chickens' Arrival

After you have set up your coop and run, and chosen your breed, number, and age of chickens, you are ready to welcome your new feathered friends to your backyard. However, before you bring them home, you need to do some preparations to ensure a smooth and successful transition. Here are some tips on how to get everything ready, introduce your chickens to their new home, and monitor their health and handle emergencies.

Getting Everything Ready

Before you pick up or receive your chickens, you need to make sure you have everything you need to care for them. Here is a checklist of the essential items and tasks you should have and do:

- *Feed:* You need to have the appropriate type and amount of feed for your chickens, depending on their age and purpose. For example, chicks need a chick starter feed, pullets need a grower feed, and layers need a layer feed. You also need to have a feeder that is clean, full, and fresh, and that is protected from rain, snow, and pests.

- *Water:* You need to have fresh, clean, and cool water for your chickens, and a waterer that is clean, full, and fresh, and that is protected from rain, snow, and pests. You also need to add some electrolytes, vitamins, or apple cider vinegar to the water to boost their immune system and prevent dehydration, especially during the first few days.

- *Bedding:* You need to have a soft, dry, and loose bedding for your coop and nesting boxes, such as straw, shavings, sand, or peat moss.

You also need to change the bedding regularly to keep it clean and dry, and to prevent diseases and parasites.

- **Heat:** You need to have a heat source for your chicks, such as a heat lamp, a heating pad, or a brooder plate. You also need to have a thermometer to monitor the temperature, and to adjust the heat source accordingly. You also need to provide some cooler areas for the chicks to escape the heat if they feel too hot.

- *Light:* You need to have a light source for your coop, such as a natural or artificial light. You also need to have a timer to control the light, and to provide 14 to 16 hours of light per day for your layers, and 8 to 10 hours of light per day for your chicks and pullets. You also need to provide some dark areas for the chickens to rest and sleep.

- *Treats:* You need to have some treats for your chickens, such as fruits, vegetables, grains, seeds, worms, or insects. You also need to limit the treats to 10% of their diet, and to provide them in moderation and variety.

You also need to avoid some toxic or unhealthy foods, such as chocolate, avocado, onion, garlic, or salt.

- *Toys:* You need to have some toys for your chickens, such as perches, swings, ladders, mirrors, balls, or bells. You also need to provide them with some natural enrichment, such as grass, weeds, herbs, flowers, or dust baths. You also need to rotate the toys and enrichment regularly to keep them interested and stimulated.

A Welcome to Your Chickens to Their New Home

After you have everything ready, you can bring your chickens to their new home. However, you need to be careful and patient when introducing them to their new environment, as they might be stressed, scared, or confused by the change. Here are some steps to follow when introducing your chickens to their new home:

- *Transport:* You need to transport your chickens in a safe, comfortable, and secure container, such as a cardboard box, a plastic crate, or a wire cage. You also need to provide them with some bedding, water, and ventilation, and to avoid extreme temperatures, noises, or movements.

You also need to transport them as quickly and smoothly as possible, and to avoid any stops or delays.

- *Quarantine:* You need to quarantine your chickens for at least two weeks before introducing them to your existing flock, if you have one. You also need to keep them in a separate coop and run, or in a separate area of your coop and run, and to prevent any direct or indirect contact with your other chickens. You also need to observe them for any signs of illness, injury, or parasites, and to treat them accordingly.

- *Acclimate:* You need to acclimate your chickens to their new coop and run, and to your existing flock, if you have one. You also need to keep them in their coop for the first few days, and to let them explore their run for the first few hours. You also need to introduce them to your other chickens gradually, and to supervise them for any signs of aggression, pecking, or bullying.

- *Bond:* You need to bond with your chickens and make them feel comfortable and happy in their new home.

You also need to handle them gently and frequently, and to offer them some treats and praise. You also need to learn their names, personalities, and preferences, and to respect their space and needs.

Monitoring Health and Handling Emergencies

Once your chickens are settled in their new home, you need to monitor their health and handle any emergencies that might arise. Chickens are generally hardy and resilient, but they can also suffer from various diseases, injuries, or predators. Here are some tips on how to monitor their health and handle emergencies:

- *Check:* You need to check your chickens daily for any signs of physical or behavioral problems, such as lethargy, loss of appetite, weight loss, diarrhea, coughing, sneezing, wheezing, limping, bleeding, or feather loss. You also need to check their coop and run daily for any signs of cleanliness, security, or damage.

- *Treat:* You need to treat your chickens promptly and properly for any minor or common issues, such as wounds, cuts, scratches, pecking, mites, lice, or worms. You also need to have a first aid kit and a medicine cabinet for your chickens, and to use some natural or home remedies, such as honey, turmeric, or garlic. You also need to isolate and quarantine any sick or injured chickens from the rest of the flock.

- *Call:* You need to call a vet or an expert for any major or rare issues, such as infections, diseases, fractures, or poisoning. You also need to have a contact list and a transport plan for your chickens, and to follow the vet's or expert's advice and instructions. You also need to euthanize or cull any terminally ill or suffering chickens, if necessary.

Part II: Caring for Your Chickens

Feeding Your Chickens

One of the most important aspects of caring for your chickens is feeding them properly. A good diet will keep your chickens healthy, happy, and productive. Here are some tips on how to provide a balanced and nutritious diet, what types and amounts of feed, treats, and supplements to give, and what best feeding practices and troubleshooting to follow.

Providing a Balanced and Nutritious Diet

Chickens are omnivorous animals, which means they can eat both plant and animal foods. They need a balanced and nutritious diet that provides them with the essential nutrients, such as protein, carbohydrates, fats, vitamins, minerals, and water. A balanced and nutritious diet will help your chickens grow, lay, and resist diseases and parasites.

49

The main components of a balanced and nutritious diet for chickens are:

- *Commercial feed:* Commercial feed is a specially formulated and processed feed that contains the optimal balance and proportion of nutrients for chickens. It comes in different forms, such as pellets, crumbles, or mash, and different types, such as starter, grower, layer, or broiler feed. Commercial feed is the easiest and most convenient way to feed your chickens, as it provides them with a complete and consistent diet. You can buy commercial feed from online or local stores, and follow the instructions on the label for how much and how often to feed your chickens.

- *Fresh food:* Fresh food is any natural and unprocessed food that you can give to your chickens as a supplement or a treat. It includes fruits, vegetables, grains, seeds, worms, insects, and meat. Fresh food is a great way to add variety, flavor, and enrichment to your chickens' diet, as it provides them with some extra nutrients, fiber, and moisture. You can grow, buy, or find fresh food for your chickens, and offer them in moderation and variety.

You can also avoid some toxic or unhealthy foods, such as chocolate, avocado, onion, garlic, or salt.

- **Grit:** Grit is any hard and insoluble material that chickens need to digest their food properly. It includes sand, gravel, oyster shells, or crushed eggshells. Grit helps the chickens grind and break down their food in their gizzard, a muscular organ in their digestive tract. Grit is especially important for chickens that eat a lot of fresh food, as it helps them digest the fiber and cellulose. You can provide grit for your chickens in a separate container, and let them eat it as they need.

Types and Amounts of Feed, Treats, and Supplements

The types and amounts of feed, treats, and supplements that you give to your chickens depend on their age, purpose, and preference. Different types of chickens have different nutritional requirements and preferences, and you need to adjust their diet accordingly.

Below are some general guidelines for the types and amounts of feed, treats, and supplements for different types of chickens:

- **Chicks:** Chicks are baby chickens that are less than 8 weeks old. They need a high-protein and high-energy diet to support their rapid growth and development. You should feed them a chick starter feed, which contains about 20% protein and 3% fat. You should also provide them with fresh water and grit, and avoid giving them any treats or supplements, unless they are prescribed by a vet or an expert. You should feed them as much as they want, and make sure they always have access to food and water.

- **Pullets:** Pullets are young female chickens that are between 8 and 20 weeks old. They need a moderate-protein and moderate-energy diet to support their continued growth and maturity. You should feed them a grower feed, which contains about 16% protein and 2.5% fat. You should also provide them with fresh water and grit, and limit giving them any treats or supplements to 10% of their diet.

You should feed them as much as they want, and make sure they always have access to food and water.

- **Layers:** Layers are adult female chickens that are older than 20 weeks and are laying eggs. They need a low-protein and high-calcium diet to support their egg production and health. You should feed them a layer feed, which contains about 14% protein and 4% calcium. You should also provide them with fresh water and grit, and limit giving them any treats or supplements to 10% of their diet. You should feed them about 1/4 to 1/3 pound of feed per day, and make sure they always have access to food and water.

- **Broilers:** Broilers are chickens that are raised for meat production. They need a high-protein and high-energy diet to support their fast growth and weight gain. You should feed them a broiler feed, which contains about 22% protein and 3.5% fat. You should also provide them with fresh water and grit, and avoid giving them any treats or supplements, unless they are prescribed by a vet or an expert.

You should feed them as much as they want, and make sure they always have access to food and water.

Age	Protein Content	Feeding Method	Feed Form
0-6 weeks	20%	Free-choice	Crumbles or mash
7-18 weeks	14-16%	Twice a day or free-choice	Pellets or mash
19+ weeks (layers)	15-18%	Twice a day or free-choice	Pellets or mash

Best Feeding Practices and Troubleshooting

To ensure the best feeding practices and troubleshoot any feeding problems, you should follow these tips:

- Use a clean, sturdy, and secure feeder that is suitable for your chickens' size and number. You can use hanging, trough, or tube feeders, and make sure they are protected from rain, snow, and pests. You should also place the feeder at the right height and distance for your chickens, and make it convenient and hygienic for them.

- Use a clean, fresh, and cool waterer that is suitable for your chickens' size and number. You can use nipple, cup, or bowl waterers, and make sure they are protected from rain, snow, and pests.

You should also place the waterer at the right height and distance for your chickens, and make it convenient and hygienic for them.

- Store your feed, treats, and supplements in a cool, dry, and dark place, and use them before their expiration date. You should also check them for any signs of mold, spoilage, or contamination, and discard them if they are bad.

- Monitor your chickens' appetite, weight, and egg production, and adjust their diet accordingly. You should also observe their behavior, droppings, and feathers, and look for any signs of illness, injury, or parasites, and treat them accordingly.

- Consult a vet or an expert if you have any questions or concerns about your chickens' diet, or if you encounter any feeding problems, such as picky eating, overeating, underfeeding, or malnutrition.

Watering Your Chickens

One of the most important aspects of caring for your chickens is watering them properly. Water is essential for your chickens' health, hydration, and egg production. Here are some tips on how to ensure a clean and fresh water supply, what types and sizes of waterers to use, and what best watering practices and problem-solving to follow.

Keeping the Water Source Fresh and Clean
Your chickens need a constant supply of clean and fresh water, especially during the hot and dry weather. Dirty or stale water can cause diseases, infections, or dehydration for your chickens. Here are some ways to ensure a clean and fresh water supply for your chickens:

- Change the water daily or more often if it gets dirty, cloudy, or frozen. You should also rinse and scrub the waterer regularly to remove any algae, dirt, or debris.

- Use a water filter, purifier, or sanitizer to remove any impurities, chemicals, or pathogens from the water.

You can also add some electrolytes, vitamins, or apple cider vinegar to the water to boost their immune system and prevent dehydration.

- Protect the water from rain, snow, and pests, such as rodents, insects, or birds. You can use a cover, a lid, or a net to shield the water from the elements and the intruders. You can also use a hanging, nipple, or cup waterer to prevent the water from spilling, splashing, or contaminating.

- Provide enough water for your chickens, depending on their size, number, and activity level. A general guideline is to provide at least one gallon of water for every four to six chickens. You should also have more than one waterer, and place them in different locations, to ensure equal access and distribution.

Types and Sizes of Waterers

There are different types and sizes of waterers that you can use for your chickens, depending on your budget, preference, and convenience.

Each type and size has its own advantages and disadvantages, and you need to choose the one that suits your situation best. Here are some of the most common types and sizes of waterers for chickens:

- **Bowl:** A bowl is a simple and cheap waterer that consists of a shallow and wide container that holds the water. It is easy to fill, clean, and move, and it allows the chickens to drink easily and naturally. However, it is also prone to spilling, splashing, and contaminating, and it can freeze or evaporate quickly. It is suitable for small and temporary flocks, or as a backup waterer.

- **Trough:** A trough is a long and narrow waterer that consists of a metal or plastic tube that holds the water. It has small holes or slots along the sides that allow the chickens to drink. It is durable, stable, and spacious, and it can hold a large amount of water. However, it is also heavy, bulky, and expensive, and it can get dirty, clogged, or frozen easily. It is suitable for large and permanent flocks, or as a main waterer.

- *Nipple:* A nipple is a modern and innovative waterer that consists of a metal or plastic valve that releases the water when the chickens peck at it. It is attached to a bucket, a bottle, or a pipe that holds the water. It is clean, efficient, and hygienic, and it prevents the water from spilling, splashing, or contaminating. However, it is also complex, fragile, and costly, and it can leak, break, or freeze easily. It is suitable for medium and advanced flocks, or as a supplementary waterer.

- *Cup:* A cup is a hybrid and versatile waterer that consists of a small and round container that holds the water. It has a trigger or a float that refills the water when the chickens drink from it. It is attached to a bucket, a bottle, or a pipe that holds the water. It is simple, reliable, and convenient, and it combines the benefits of the bowl and the nipple waterers. However, it is also limited, variable, and pricey, and it can overflow, jam, or freeze easily. It is suitable for any size and level of flocks, or as an alternative waterer.

Best Watering Practices and Problem-solving

To ensure the best watering practices and problem-solving, you should follow these tips:

- Use a clean, sturdy, and secure waterer that is suitable for your chickens' size and number. You can use any type and size of waterer, and make sure they are protected from rain, snow, and pests. You should also place the waterer at the right height and distance for your chickens, and make it convenient and hygienic for them.

- Use fresh, clean, and cool water for your chickens, and change it daily or more often if it gets dirty, cloudy, or frozen. You can also use a water filter, purifier, or sanitizer to remove any impurities, chemicals, or pathogens from the water. You can also add some electrolytes, vitamins, or apple cider vinegar to the water to boost their immune system and prevent dehydration.

- Monitor your chickens' water intake, and adjust it accordingly. You should also observe their behavior, droppings, and feathers, and look for any signs of thirst, dehydration, or disease, and treat them accordingly.

- Consult a vet or an expert if you have any questions or concerns about your chickens' water, or if you encounter any watering problems, such as leaking, freezing, or poisoning.

Cleaning Your Chickens and Their Coop

One of the most important aspects of caring for your chickens is cleaning them and their coop. A clean and hygienic environment will keep your chickens healthy, happy, and productive. Here are some tips on how to maintain cleanliness and hygiene, what cleaning methods and tools to use, and what bedding choices and maintenance to follow.

Maintaining Cleanliness and Hygiene

Your chickens and their coop need to be cleaned regularly to prevent diseases, parasites, and odors. Dirty or damp conditions can cause infections, respiratory problems, or fungal growth for your chickens.

The following are some ways to maintain cleanliness and hygiene for your chickens and their coop:

- Remove any droppings, feathers, or debris from the coop and the run daily or more often if needed. You can use a rake, a shovel, or a scraper to collect and dispose of the waste. You can also use the waste as a source of compost or fertilizer for your garden.

- Change the bedding in the coop and the nesting boxes weekly or more often if needed. You can use materials such as straw, shavings, sand, or peat moss as bedding. You should also fluff and rotate the bedding regularly to keep it dry and loose.

- Scrub and disinfect the coop and the run monthly or more often if needed. You can use a hose, a brush, or a sponge to wash and rinse the coop and the run. You can also use a natural or commercial cleaner to sanitize and deodorize the coop and the run. You should also let the coop and the run dry completely before adding new bedding or letting the chickens back in.

- Dust and treat the chickens and the coop for any mites, lice, or worms quarterly or more often if needed. You can use a natural or commercial dust, spray, or powder to kill and repel the parasites. You can also use a natural or commercial wormer to eliminate and prevent the worms. You should also follow the instructions on the label for how much and how often to apply the products.

Cleaning Methods and Tools

There are different cleaning methods and tools that you can use for your chickens and their coop, depending on your budget, preference, and convenience. Each method and tool has its own advantages and disadvantages, and you need to choose the one that suits your situation best. Below are some of the most common cleaning methods and tools for chickens and their coop:

- *Deep litter:* Deep litter is a cleaning method that involves adding fresh bedding on top of the old bedding, instead of removing and replacing it. The old bedding decomposes and creates heat and compost for the chickens. The fresh bedding absorbs and covers the droppings and the odor.

The deep litter method is easy, cheap, and beneficial, as it provides insulation, fertilizer, and enrichment for the chickens. However, it is also risky, messy, and smelly, as it can harbor diseases, parasites, or ammonia if not managed properly. It is suitable for cold and dry climates, and for large and well-ventilated coops.

- *Spot cleaning:* Spot cleaning is a cleaning method that involves removing and replacing only the dirty or wet spots of the bedding, instead of the whole bedding. The clean bedding remains and saves the waste and the cost. The spot cleaning method is quick, simple, and economical, as it reduces the amount of bedding and labor needed. However, it is also incomplete, inconsistent, and tedious, as it can miss some hidden or scattered droppings or debris. It is suitable for warm and humid climates, and for small and easy-to-access coops.

- **Complete cleaning:** Complete cleaning is a cleaning method that involves removing and replacing the entire bedding, and washing and disinfecting the coop and the run. The old bedding is discarded and the new bedding is added.

The coop and the run are scrubbed and sanitized. The complete cleaning method is thorough, effective, and hygienic, as it eliminates all the dirt, germs, and smells. However, it is also time-consuming, labor-intensive, and wasteful, as it requires a lot of bedding, water, and cleaner. It is suitable for any climate and coop, but it should be done less frequently than the other methods.

- *Rake:* A rake is a cleaning tool that consists of a long handle and a metal or plastic head with teeth or tines. It is used to collect and remove the droppings, feathers, or debris from the coop and the run. It is cheap, easy, and versatile, and it can reach and cover a large area. However, it is also rough, loose, and inefficient, and it can miss or scatter some of the waste. It is suitable for any type of bedding, but it should be used with caution and care.

- *Shovel:* A shovel is a cleaning tool that consists of a long handle and a metal or plastic blade with a flat or curved edge. It is used to scoop and dispose of the bedding, droppings, or debris from the coop and the run. It is sturdy, stable, and effective, and it can lift and carry a large amount of waste. However, it is also heavy, bulky, and awkward, and it can damage or disturb the coop or the run. It is suitable for any type of bedding, but it should be used with skill and strength.

- *Scraper:* A scraper is a cleaning tool that consists of a short handle and a metal or plastic blade with a sharp or serrated edge. It is used to scrape and loosen the droppings, feathers, or debris from the coop and the run. It is light, handy, and precise, and it can reach and remove the stubborn or stuck waste. However, it is also fragile, limited, and tedious, and it can scratch or injure the coop or the run. It is suitable for any type of bedding, but it should be used with patience and caution.

Bedding Choices and Maintenance

The bedding is the material that covers the floor and the nesting boxes of the coop. It provides comfort, warmth, and cushion for the chickens. It also absorbs and masks the droppings, moisture, and odor. The bedding needs to be chosen and maintained carefully to ensure the cleanliness and hygiene of the coop and the chickens.

Consider the following tips on how to choose and maintain the bedding for your coop:

- Choose a bedding that is soft, dry, and loose, and that suits your climate, coop, and budget. You can use materials such as straw, shavings, sand, or peat moss as bedding. You should also avoid materials such as hay, newspaper, or sawdust, as they can be dusty, moldy, or slippery.

- Choose a bedding that is easy to find, buy, or make, and that is eco-friendly and biodegradable. You can use materials such as straw, shavings, sand, or peat moss as bedding. You should also avoid materials such as plastic, rubber, or synthetic fibers, as they can be harmful, expensive, or nonrenewable.

- Choose a bedding that is compatible, attractive, and beneficial for your chickens. You can use materials such as straw, shavings, sand, or peat moss as bedding. You should also avoid materials such as cedar, pine, or walnut, as they can be toxic, irritating, or allergic for your chickens.

- Maintain the bedding by changing it regularly, depending on the type, amount, and condition of the bedding. You can use the deep litter, spot cleaning, or complete cleaning methods to change the bedding. You should also fluff and rotate the bedding regularly to keep it dry and loose.

- Maintain the bedding by using it as a source of compost or fertilizer for your garden, or as a fuel or bedding for your other animals. You can also donate or sell the bedding to your neighbors, friends, or local farms. You should also dispose of the bedding properly and safely, and follow the local laws and regulations regarding the waste management.

Protecting Your Chickens

One of the most important aspects of caring for your chickens is protecting them from any harm or danger. Chickens can face various threats, such as diseases, parasites, injuries, predators, or pests. Here are some tips on how to prevent and treat diseases, parasites, and injuries, how to recognize signs of illness and basic first aid, and how to prevent predators and pests.

Preventing and Treating Diseases, Parasites, and Injuries

Chickens can suffer from various diseases, parasites, and injuries, which can affect their health, productivity, and happiness. Some of the most common ones are:

- **The Diseases:** Diseases are illnesses caused by bacteria, viruses, fungi, or protozoa. They can affect the respiratory, digestive, nervous, or reproductive systems of the chickens. Some of the most common diseases are coccidiosis, infectious bronchitis, Marek's disease, Newcastle disease, and fowl pox.

- **Parasites:** Parasites are organisms that live on or in the chickens and feed on their blood, skin, or feathers. They can cause irritation, inflammation, infection, or anemia for the chickens. Some of the most common parasites are mites, lice, worms, and fleas.

- **Injuries:** Injuries are wounds or damages caused by accidents, fights, or predators. They can affect the skin, bones, muscles, or organs of the chickens.

Some of the most common injuries are cuts, scratches, pecking, fractures, or bites.

To prevent and treat diseases, parasites, and injuries, you should follow these steps

- *Prevention:* Prevention is the best way to protect your chickens from diseases, parasites, and injuries. You should provide them with a clean, comfortable, and secure coop and run, a balanced and nutritious diet, and fresh and clean water. You should also vaccinate, deworm, and dust them regularly, and quarantine and isolate any new or sick chickens from the rest of the flock. You should also monitor and check your chickens daily for any signs of physical or behavioral problems, and consult a vet or an expert if you have any questions or concerns.

- *Treatment:* Treatment is the next best way to protect your chickens from diseases, parasites, and injuries. You should have a first aid kit and a medicine cabinet for your chickens, and use some natural or commercial remedies, such as honey, turmeric, garlic, or antibiotics. You should also follow the vet's or expert's advice and instructions, and treat your chickens promptly and properly. You should also isolate and quarantine any sick or injured chickens from the rest of the flock, and euthanize or cull any terminally ill or suffering chickens, if necessary.

Recognizing Signs of Illness and Basic First Aid

Chickens can show various signs of illness, which can indicate that they are suffering from diseases, parasites, or injuries. Some of the most common signs of illness are:

- *Physical signs:* Physical signs are changes in the appearance or condition of the chickens. They include lethargy, loss of appetite, weight loss, diarrhea, coughing, sneezing, wheezing, limping, bleeding, or feather loss.

- *Behavioral signs:* Behavioral signs are changes in the attitude or activity of the chickens. They include depression, isolation, aggression, nervousness, or abnormal vocalization.

- *Egg signs:* Egg signs are changes in the quantity or quality of the eggs. They include reduced or increased egg production, abnormal egg shape, size, color, or shell, or blood or mucus in the eggs.

To recognize signs of illness, you should observe your chickens closely and regularly, and compare them with their normal or healthy state. You should also use a thermometer, a scale, or a candling device to measure their temperature, weight, or egg development. You should also keep a record or a chart of their health and performance, and note any changes or abnormalities.

To provide basic first aid, you should follow these steps:
- *Assess:* Assess the situation and the severity of the problem. You should identify the cause, type, and location of the problem,

and determine if it is minor or major, common or rare, or acute or chronic. You should also check the vital signs of the chicken, such as breathing, heartbeat, and temperature, and see if they are normal or abnormal.

- **Stabilize:** Stabilize the condition and the comfort of the chicken. You should stop any bleeding, clean any wound, apply any bandage, splint, or dressing, and administer any painkiller, anti-inflammatory, or antiseptic. You should also provide some water, electrolytes, or vitamins, and move the chicken to a safe, warm, and quiet place.

- **Treat:** Treat the problem and the symptoms of the chicken. You should use some natural or commercial remedies, such as honey, turmeric, garlic, or antibiotics, and follow the instructions on the label for how much and how often to apply them. You should also follow the vet's or expert's advice and instructions, and monitor the progress and recovery of the chicken.

Predator and Pest Prevention

Chickens can face various threats from predators and pests, which can harm or kill them. Predators are animals that hunt and attack chickens, such as dogs, cats, foxes, raccoons, hawks, or snakes. Pests are animals that annoy and bother chickens, such as rodents, insects, or birds. To prevent predators and pests, you should follow these steps:

- *Secure:* Secure your coop and run from predators and pests. You should use materials such as wire, mesh, or electric fencing, and hardware, locks, or latches, to enclose and protect your coop and run. You should also check your coop and run regularly for any gaps, holes, or weak points, and repair them as soon as possible.

- *Deter:* Deter predators and pests from your coop and run. You can use methods such as noise, light, or motion, to scare and repel predators and pests. You can also use methods such as traps, repellents, or decoys, to catch and remove predators and pests. You should also keep your coop and run clean and tidy, and remove any food, water, or waste that might attract predators and pests.

- **_Defend:_** Defend your chickens from predators and pests. You can use methods such as dogs, roosters, or guinea fowl, to guard and alert your chickens. You can also use methods such as wings, spurs, or beaks, to arm and equip your chickens. You should also supervise and watch your chickens, especially when they are free-ranging, and intervene and assist them if needed.

Part III: Enjoying Your Chickens

Collecting and Storing Your Eggs

One of the most rewarding aspects of raising backyard chickens is collecting and storing your eggs. Fresh eggs from your own chickens are delicious, nutritious, and satisfying. Here are some tips on how to harvest and handle eggs, what factors affect egg quality and quantity, and what best practices for egg collection and storage to follow.

Harvesting and Handling Eggs

Harvesting and handling eggs are the processes of gathering and preparing your eggs for consumption or preservation. You need to be careful and gentle when harvesting and handling eggs, as they are fragile and perishable. Here are some steps to follow when harvesting and handling eggs:

- **Collect:** Collect your eggs from the nesting boxes daily or more often if needed. You can use a basket, a bucket, or a carton to hold and carry your eggs. You should also wear gloves, wash your hands, or sanitize your tools to prevent any contamination or infection.

- *Inspect:* Inspect your eggs for any cracks, dirt, or defects. You can use a candling device, a flashlight, or a bright light to check the inside and the outside of your eggs. You should also weigh, measure, or grade your eggs according to their size, shape, color, or quality.

- *Clean:* Clean your eggs if they are dirty or stained. You can use a dry or a wet method to clean your eggs. The dry method involves using a brush, a cloth, or a sandpaper to gently rub and remove the dirt from your eggs. The wet method involves using water, soap, or vinegar to gently wash and rinse your eggs. You should also dry your eggs thoroughly after cleaning them.

- *Store:* Store your eggs in a cool, dry, and dark place, such as a refrigerator, a cellar, or a pantry. You can use a carton, a tray, or a rack to store your eggs. You should also label, date, or rotate your eggs according to their freshness, type, or origin.

Factors Affecting Egg Quality and Quantity

Egg quality and quantity are the measures of the appearance, condition, and performance of your eggs. They depend on various factors, such as your chickens' breed, age, diet, health, and environment. Here are some of the most important factors affecting egg quality and quantity:

- *Breed:* Breed is the genetic factor that determines the potential and the characteristics of your chickens and their eggs. Different breeds have different egg production, size, color, and shape. For example, Leghorns are known for their high egg production and white eggs, while Orpingtons are known for their large eggs and buff color.

- *Age:* Age is the physiological factor that affects the maturity and the productivity of your chickens and their eggs. Younger chickens tend to lay smaller, fewer, and irregular eggs, while older chickens tend to lay larger, more, and consistent eggs. However, as chickens age, their egg production and quality decline gradually.

- **Diet:** Diet is the nutritional factor that influences the health and the performance of your chickens and their eggs. A balanced and nutritious diet will provide your chickens with the essential nutrients, such as protein, calcium, and vitamins, to produce healthy and quality eggs. A poor or inadequate diet will cause your chickens to produce unhealthy and low-quality eggs.

- **Health:** Health is the physical factor that affects the well-being and the efficiency of your chickens and their eggs. Healthy chickens will produce more and better eggs, while sick or injured chickens will produce less and worse eggs. Diseases, parasites, or injuries can also affect the appearance, condition, or safety of your eggs.

- **Environment:** Environment is the external factor that impacts the comfort and the behavior of your chickens and their eggs.

A comfortable and stimulating environment will encourage your chickens to lay more and better eggs, while a stressful or harsh environment will discourage your chickens to lay less and worse eggs. Temperature, light, space, and security are some of the environmental factors that affect your chickens and their eggs.

Best Practices for Egg Collection and Storage

To ensure the best practices for egg collection and storage, you should follow these tips:

- Collect your eggs daily or more often if needed, and inspect them for any cracks, dirt, or defects. You should also discard any bad or questionable eggs, and keep only the good and fresh ones.

- Clean your eggs if they are dirty or stained, using a dry or a wet method. You should also dry your eggs thoroughly after cleaning them, and avoid washing them before storing them, as it can remove the natural protective coating or bloom from the eggs.

- Store your eggs in a cool, dry, and dark place, such as a refrigerator, a cellar, or a pantry. You should also store your eggs in a carton, a tray, or a rack, and label, date, or rotate them according to their freshness, type, or origin.

- Use your eggs within three to five weeks of collecting them, or longer if you preserve them by freezing, pickling, or dehydrating them. You should also use your eggs within a week of washing them, or sooner if you crack them. You should also use your eggs within a day of cooking them, or sooner if you serve them raw or undercooked.

Interacting with Your Chickens

Interacting with your chickens is a rewarding and enjoyable way to enhance your relationship with them. Chickens are intelligent, social, and curious animals that can benefit from your attention and affection. Here are some tips on how to bond, handle, and train your chickens, how to create toys and treats for them, and how to have fun and relax with your feathered friends.

Bonding, Handling, and Training

Bonding, handling, and training are the processes of establishing and strengthening your connection and communication with your chickens. You can bond, handle, and train your chickens to make them more friendly, tame, and obedient. Here are some steps to follow when bonding, handling, and training your chickens:

- *Start early:* The best time to bond, handle, and train your chickens is when they are young and impressionable. Chicks are more receptive and adaptable to human interaction than adult chickens. You can start bonding, handling, and training your chicks as soon as they hatch, or as soon as you bring them home.

- *Be gentle:* The most important thing to bond, handle, and train your chickens is to be gentle and kind to them. You should avoid any harsh, sudden, or loud actions or noises that might scare or hurt your chickens. You should also respect their space and needs, and not force them to do anything they don't want to do.

- **Be consistent:** The key to bond, handle, and train your chickens is to be consistent and persistent with them. You should interact with your chickens regularly and frequently, and follow a routine and a schedule. You should also use the same words, gestures, and signals to communicate with your chickens, and reward or correct them accordingly.

- **Be patient:** The challenge to bond, handle, and train your chickens is to be patient and realistic with them. You should understand that chickens have different personalities, preferences, and abilities, and that they learn at different paces. You should also appreciate that chickens are not dogs or cats, and that they have their own limits and instincts.

Creating Toys and Treats for Chickens

Creating toys and treats for your chickens is a fun and creative way to enrich and stimulate their lives. Chickens are playful, curious, and hungry animals that can enjoy and benefit from your toys and treats.

Below are some ideas on how to create toys and treats for your chickens:

- **Toys:** Toys are any objects or items that you can give to your chickens to entertain and challenge them. You can use materials such as wood, metal, plastic, or recycled materials, and make them into perches, swings, ladders, mirrors, balls, or bells. You can also use natural materials such as grass, weeds, herbs, flowers, or dust baths, and make them into foraging, scratching, or dusting areas. You should also rotate and replace the toys regularly to keep them interesting and fresh for your chickens.

- **Treats:** Treats are any foods or snacks that you can give to your chickens to reward and satisfy them. You can use fruits, vegetables, grains, seeds, worms, insects, or meat, and offer them in moderation and variety. You can also use yogurt, cheese, eggs, or bread, and make them into cakes, muffins, or cookies. You should also limit the treats to 10% of their diet, and avoid any toxic or unhealthy foods, such as chocolate, avocado, onion, garlic, or salt.

Fun and Relaxation with Your Feathered Friends

Having fun and relaxation with your feathered friends is a relaxing and enjoyable way to spend your time and energy with them. Chickens are amusing, amusing, and amusing animals that can make you laugh and smile. Let's walk you through some ways to have fun and relax with your feathered friends:

- *Watch:* Watching your chickens is one of the simplest and easiest ways to have fun and relax with them. You can observe their behavior, interactions, and expressions, and learn more about their personalities, moods, and preferences. You can also admire their beauty, colors, and patterns, and appreciate their diversity, uniqueness, and charm.

- *Talk:* Talking to your chickens is one of the most natural and common ways to have fun and relax with them. You can use your voice, tone, and words to communicate and bond with your chickens.

- You can also use your body, face, and eyes to express and convey your emotions and feelings to your chickens. You can also listen to their sounds, clucks, and purrs, and understand their messages and signals to you.

- *Cuddle:* Cuddling with your chickens is one of the most intimate and affectionate ways to have fun and relax with them. You can use your hands, arms, and lap to hold and hug your chickens. You can also use your fingers, nails, and lips to pet and kiss your chickens. You can also feel their warmth, softness, and heartbeat, and share your comfort and love with your chickens.

Using Your Chickens' Manure

If you keep chickens in your backyard or farm, you may wonder what to do with the large amount of manure they produce. Chicken manure is not just a waste product, but a valuable resource that can benefit your garden, your soil, and your environment. Here are some tips on how to turn manure into a valuable resource, what benefits and uses of chicken manure are, and what management, processing, and environmental precautions to follow.

Turning Manure into a Valuable Resource

Chicken manure is a rich source of nutrients, organic matter, and microorganisms that can improve the fertility, structure, and health of your soil. However, fresh chicken manure is too strong and too hot to be used directly on your plants, as it can burn them or cause diseases. Therefore, you need to turn manure into a valuable resource by composting or aging it before you use it.

Composting is the process of breaking down organic materials into a dark, crumbly, and odorless substance called compost. Compost is a natural fertilizer that can provide your plants with the essential nutrients they need to grow and thrive. To compost chicken manure, you need to mix it with other organic materials, such as leaves, grass clippings, straw, or wood chips, in a ratio of 1 part manure to 2 parts carbon-rich materials. You also need to add some water, air, and heat to the mixture, and turn it regularly to speed up the decomposition. The composting process can take from a few weeks to a few months, depending on the conditions and the materials. You can tell when the compost is ready by its appearance, smell, and temperature. It should look like dark soil, smell earthy, and feel cool to the touch.

Aging is the process of letting organic materials sit and rot naturally over time. Aging is a simpler and slower method than composting, but it can also produce a useful fertilizer for your plants. To age chicken manure, you need to pile it up in a shady and dry place, and cover it with a tarp or a plastic sheet to prevent rain or pests from getting in. You also need to let it sit for at least six months to a year, or until it loses its strong smell and color. You can tell when the aged manure is ready by its texture, smell, and pH. It should be crumbly, odorless, and neutral.

Benefits and Uses of Chicken Manure

Chicken manure has many benefits and uses for your garden, your soil, and your environment. Consider the following:

- *Benefits:* Chicken manure is a complete fertilizer that contains the macronutrients nitrogen, phosphorus, and potassium, as well as important micronutrients such as calcium, magnesium, and sulfur. Your plants can grow more robustly, quickly, and healthily with the aid of these nutrients. Chicken manure is also a good soil amendment that can improve the texture, drainage, and water retention of your soil. It can also increase the organic matter and the microbial activity of your soil, which can enhance the soil's fertility and resilience. Chicken manure can also help reduce the need for synthetic fertilizers, which can be harmful to the environment and your health. Chicken manure can also help reduce the amount of waste that goes to the landfill, which can reduce greenhouse gas emissions and pollution.

- *Uses:* Chicken manure can be used in various ways for your garden, depending on your needs and preferences. You can use it as a top dressing, a side dressing, or a mulch for your plants. You can also use it as a base, a component, or a supplement for your compost. You can also use it as a tea, a liquid, or a foliar fertilizer for your plants. You can also use it as a soil conditioner, a starter, or a booster for your soil. You can also use it as a pest repellent, a disease fighter, or a weed killer for your garden.

Management, Processing, and Environmental Precautions

Chicken manure is a valuable resource, but it also requires proper management, processing, and environmental precautions to ensure its safety and effectiveness. Consider the tips below:

- *Management:* You need to manage your chicken manure by collecting, storing, and disposing of it properly. You need to collect your chicken manure daily or more often if needed, and inspect it for any signs of mold, spoilage, or contamination. You need to store your chicken manure in a cool, dry, and dark place, and use it before its expiration date. You need to dispose of your chicken manure properly and safely, and follow the local laws and regulations regarding the waste management.

- *Processing:* You need to process your chicken manure by composting or aging it before you use it. You need to compost your chicken manure by mixing it with other organic materials, adding some water, air, and heat, and turning it regularly. You need to age your chicken manure by piling it up in a shady and dry place, and covering it with a tarp or a plastic sheet. You need to check your chicken manure for its readiness by its appearance, smell, and temperature.

- *Environmental Precautions:* You need to take some environmental precautions when using your chicken manure to prevent any negative impacts on the environment and your health. You need to use your chicken manure in moderation and according to your plants' needs, and avoid over-fertilizing or under-fertilizing your plants. You need to use your chicken manure in the right time and place, and avoid applying it near water sources or sensitive areas. You need to use your chicken manure in the right form and method, and avoid spreading it raw or unprocessed on your plants.

Part IV: Learning More About Your Chickens

Understanding Your Chickens' Anatomy and Physiology

If you want to learn more about your chickens, you need to understand their anatomy and physiology. The study of anatomy focuses on the composition and form of the body, whereas the study of physiology examines the functions and processes of the body and its components. By understanding your chickens' anatomy and physiology, you can appreciate their uniqueness, complexity, and diversity. Here are some tips on how to appreciate chicken anatomy and major organ systems, how to learn the basics of chicken physiology, and how to compare chickens to other animals.

Appreciating Chicken Anatomy and Major Organ Systems

Chicken anatomy is the physical form and structure of the chicken's body and its parts. It includes the external and internal features, such as the feathers, skin, bones, muscles, organs, and systems. Chicken anatomy is adapted to suit the chicken's lifestyle, behavior, and environment. Here are some of the major organ systems of the chicken and their functions:

- *Skeletal system:* The skeletal system is the framework of bones and cartilages that supports and protects the body and its organs. It also provides attachment for the muscles and enables movement. The chicken's skeletal system is light, hollow, and flexible, which allows the chicken to fly, walk, and perch. The chicken's skeletal system also has some unique features, such as the keel, a large breastbone that anchors the flight muscles, and the pygostyle, a fused tailbone that supports the tail feathers.

- *Muscular system:* The muscular system is the group of tissues that contracts and relaxes to produce movement and force.

It also helps maintain posture and body temperature. The chicken's muscular system is divided into two types: the skeletal muscles, which are attached to the bones and are under voluntary control, and the smooth muscles, which are found in the organs and are under involuntary control. The chicken's muscular system is specialized for flight, digestion, and reproduction. For example, the pectoralis major and the supracoracoideus are the main flight muscles that move the wings, while the gizzard is a muscular organ that grinds the food, and the oviduct is a muscular tube that forms and lays the eggs.

- **Digestive system:** The digestive system is the group of organs that breaks down food into nutrients and eliminates waste. It includes the mouth, esophagus, crop, proventriculus, gizzard, small intestine, large intestine, ceca, and cloaca. The chicken's digestive system is different from the human digestive system in several ways. For instance, the chicken does not have teeth, but uses the beak and the tongue to peck and swallow the food. The chicken also does not have a stomach, but has two chambers: the proventriculus, which secretes digestive juices, and the gizzard, which crushes the food with the help of grit. The chicken also has two blind pouches called ceca, which harbor beneficial bacteria and aid in fermentation.

- *Respiratory system:* The respiratory system is the group of organs that exchanges gases between the body and the environment. It includes the nostrils, trachea, bronchi, lungs, and air sacs. The chicken's respiratory system is more efficient and complex than the human respiratory system. For example, the chicken has nine air sacs that extend into the bones and the abdominal cavity, which allow the chicken to breathe continuously and store extra air. The chicken also has a syrinx, a vocal organ at the base of the trachea, which produces sounds and vocalizations.

- *Circulatory system:* The circulatory system is the group of organs that transports blood, oxygen, nutrients, hormones, and waste throughout the body. It includes the heart, arteries, veins, and capillaries. The chicken's circulatory system is similar to the human circulatory system, but has some differences.

For instance, the chicken has a four-chambered heart, like humans, but the heart is located more to the right than the center of the chest. The chicken also has a higher heart rate and blood pressure than humans, which helps the chicken cope with the demands of flight and metabolism.

- *Nervous system:* The nervous system is the group of organs that coordinates and controls the activities of the body and its parts. It includes the brain, spinal cord, nerves, and sense organs. The chicken's nervous system is comparable to the human nervous system, but has some variations. For example, the chicken has a smaller and simpler brain than humans, but has a larger and more developed cerebellum, which controls balance and coordination. The chicken also has a better sense of sight, hearing, and smell than humans, but a poorer sense of taste and touch.

- *Reproductive system:* The reproductive system is the group of organs that produces and delivers gametes, or sex cells, and enables sexual reproduction. It includes the testes, vas deferens, cloaca, and phallus in males, and the ovaries, oviduct, cloaca, and vent in females. The chicken's reproductive system is different from the human reproductive system in several ways. For example, the male chicken does not have a penis, but a small organ called the phallus, which transfers sperm to the female's cloaca. The female chicken does not have a uterus, but a single functional ovary and oviduct, which produces and lays one egg per day.

Basics of Chicken Physiology

Chicken physiology is the study of how the chicken's body and its parts work and function. It includes the processes and mechanisms that regulate and maintain the chicken's life, such as metabolism, homeostasis, and reproduction. Chicken physiology is influenced by various factors, such as genetics, environment, and nutrition. Here are some of the basics of chicken physiology:

- *Metabolism:* Metabolism is the process of converting food into energy and building blocks for the body. It involves two types of reactions: catabolism, which breaks down molecules and releases energy, and anabolism, which builds up molecules and consumes energy.

The chicken's metabolism is faster and higher than the human metabolism, which means the chicken needs more food and oxygen to sustain its body functions. The chicken's metabolism is also affected by factors such as age, breed, activity, and temperature. For example, younger, smaller, and more active chickens have higher metabolic rates than older, larger, and less active chickens. Chickens also have higher metabolic rates in colder temperatures than in warmer temperatures, as they need more energy to keep warm.

- *Homeostasis:* Homeostasis is the process of maintaining a stable and balanced internal environment for the body. It involves various systems and organs that monitor and adjust the levels of temperature, pH, blood pressure, blood sugar, hormones, and other variables. The chicken's homeostasis is similar to the human homeostasis, but has some adaptations and challenges. For example, the chicken has a higher body temperature than humans, ranging from 105°F to 107°F, which helps the chicken fight infections and support metabolism.

However, the chicken also has a harder time regulating its body temperature, as it lacks sweat glands and has feathers that trap heat. Therefore, the chicken relies on other methods, such as panting, fluffing, or shivering, to cool or warm itself.

• **Reproduction:** Reproduction is the process of producing offspring and ensuring the continuity of the species. It involves the formation and fusion of gametes, or sex cells, and the development and delivery of embryos, or fertilized eggs. The chicken's reproduction is different from the human reproduction in several ways. For example, the chicken has a shorter and simpler reproductive cycle than humans, lasting about 25 to 26 hours, which consists of ovulation, fertilization, and oviposition. The chicken also has a different and diverse reproductive system than humans, as described in the previous section. The chicken also has a different and variable reproductive output than humans, as it can produce and lay up to 300 eggs per year, depending on the breed, age, season, and management.

Comparing Chickens to Other Animals

Chickens are animals that belong to the class Aves, or birds, which is a group of vertebrates that have feathers, wings, and a beak. Chickens are also members of the order Galliformes, or fowls, which is a group of ground-dwelling birds that have a heavy body, short legs, and strong feet. Chickens are also part of the family Phasianidae, or pheasants, which is a group of colorful and diverse birds that have a long tail, a crest, and a wattle. Chickens are also part of the genus Gallus, or junglefowl, which is a group of four wild species that are the ancestors of domestic chickens.

Chickens are similar to other animals in some ways, but different in others. Here are some of the similarities and differences between chickens and other animals:

- *Similarities:* Chickens share some common characteristics and features with other animals, such as having a backbone, a heart, a brain, a liver, a stomach, and a pair of lungs. Chickens also share some common behaviors and functions with other animals, such as eating, drinking, breathing, sleeping, and reproducing. Chickens also share some common ancestry and evolution with other animals, such as being descended from reptiles, having a common ancestor with mammals, and being related to dinosaurs.

- *Differences:* Chickens have some unique characteristics and features that distinguish them from other animals, such as having feathers, wings, and a beak, which enable them to fly, walk, and peck. Chickens also have some unique behaviors and functions that separate them from other animals, such as laying eggs, clucking, and dust bathing, which help them produce offspring communicate, and protect themselves. Chickens also have some unique anatomy and physiology that enable them to adapt and survive in various environments and conditions.

Exploring Your Chickens' Behavior and Intelligence

Chickens are not just mindless and noisy animals that lay eggs and peck at the ground. They are actually intelligent, social, and complex creatures that have their own behavior and communication patterns. By exploring your chickens' behavior and intelligence, you can learn more about their personalities, needs, and preferences. Here are some tips on how to observe behavior and communication, how to understand common behaviors and influencing factors, and how to demonstrate chicken intelligence and learning abilities.

Observing Behavior and Communication

Observing behavior and communication is the process of watching and listening to your chickens and noticing their actions and sounds. You can observe behavior and communication to understand what your chickens are doing, feeling, and saying.

The following are steps to follow when observing behavior and communication:

- *Set up:* Set up a comfortable and convenient spot to observe your chickens. You can use a chair, a bench, or a stool to sit and watch your chickens. You can also use a camera, a recorder, or a notebook to capture and record your chickens' behavior and communication. You should also choose a time and place that is suitable and natural for your chickens, such as the morning, the evening, or the coop, the run, or the yard.

- *Watch:* Watch your chickens closely and carefully, and pay attention to their movements, expressions, and interactions. You can look for cues such as their posture, their eyes, their beak, their wings, their tail, and their feathers, and see how they change according to their mood, state, or intention. You can also look for patterns such as their routine, their habits, and their preferences, and see how they vary according to their personality, role, or situation.

- *Listen:* Listen to your chickens attentively and curiously, and hear their sounds, clucks, and purrs.

You can identify different types of vocalizations, such as alarm, food, mating, or social calls, and see how they convey different messages, such as danger, excitement, attraction, or affiliation. You can also identify different tones, pitches, and volumes, and see how they reflect different emotions, such as fear, joy, desire, or affection.

Understanding Common Behaviors and Influencing Factors

Understanding common behaviors and influencing factors is the process of learning and explaining the reasons and meanings behind your chickens' behavior and communication. You can understand common behaviors and influencing factors to appreciate and respect your chickens' needs and choices. Here are some of the common behaviors and influencing factors of chickens and their functions:

- *Dust bathing:* Dust bathing is a behavior where chickens roll and rub themselves in the dirt, sand, or ash. It helps them clean their feathers, skin, and parasites, and also regulate their body temperature and moisture.

Dust bathing is influenced by factors such as availability, quality, and quantity of dust, as well as weather, season, and time of day.

- **_Foraging:_** Foraging is a behavior where chickens scratch and peck at the ground, looking for food, such as seeds, insects, or worms. It helps them satisfy their hunger, curiosity, and boredom, and also stimulate their senses and muscles. Foraging is influenced by factors such as availability, variety, and quantity of food, as well as competition, cooperation, and hierarchy among chickens.

- **_Roosting:_** Roosting is a behavior where chickens perch and sleep on elevated places, such as branches, poles, or rafters. It helps them feel safe, comfortable, and warm, and also establish their social order and position. Roosting is influenced by factors such as availability, quality, and quantity of roosts, as well as light, temperature, and noise levels.

Demonstrating Chicken Intelligence and Learning Abilities

Demonstrating chicken intelligence and learning abilities is the process of testing and showing your chickens' cognitive and mental skills and capacities. You can demonstrate chicken intelligence and learning abilities to challenge and enrich your chickens' lives and minds. Consider the below ways to explore chicken intelligence and learning abilities:

- *Puzzles:* Puzzles are tasks or games that require your chickens to solve problems, find solutions, or achieve goals. You can use objects or items such as boxes, containers, or toys, and hide or place food, treats, or rewards inside or behind them. You can also use shapes, colors, or symbols, and associate them with food, treats, or rewards. You can then observe how your chickens use their logic, memory, or recognition to access or obtain the food, treats, or rewards.

- **Tricks:** Tricks are actions or behaviors that your chickens can perform on command or cue. You can use words, gestures, or signals to communicate and instruct your chickens. You can also use food, treats, or rewards to motivate and reinforce your chickens. You can then teach your chickens to do simple or complex tricks, such as come, sit, stay, jump, or fetch.

- **Games:** Games are activities or competitions that involve your chickens' participation, interaction, or cooperation. You can use objects or items such as balls, ropes, or rings, and encourage your chickens to play with them. You can also use rules, scores, or teams, and challenge your chickens to follow them. You can then enjoy how your chickens use their agility, coordination, or teamwork to play the games.

Discovering Your Chickens' History and Evolution

Chickens are one of the most common and widespread domestic animals in the world. They are also one of the most diverse and fascinating animals, with a long and complex history and evolution. By discovering your chickens' history and evolution, you can learn more about their origins, development, and diversity. Here are some tips on how to trace the origins and development of chickens, how to understand the history and evolution of chickens, and how to appreciate the genetic diversity and cultural significance of chickens.

Tracing the Origins and Development of Chickens

The origins and development of chickens are the processes of how chickens came to be and how they changed over time. They involve the genetic, geographic, and cultural factors that influenced the emergence and transformation of chickens. Follow the steps below to trace the origins and development of chickens:

- *Identify the ancestors:* The first step to trace the origins and development of chickens is to identify their ancestors, or the wild species that gave rise to domestic chickens. Most scientists agree that the primary ancestor of chickens is the red junglefowl (Gallus gallus), a bird that lives in the forests of Southeast Asia. However, some scientists also suggest that other junglefowl species, such as the grey junglefowl (Gallus sonneratii), the green junglefowl (Gallus varius), and the Ceylon junglefowl (Gallus lafayettii), may have contributed to the chicken's gene pool.

- **Locate the regions:** The next step to trace the origins and development of chickens is to locate the regions, or the places where chickens were first domesticated and spread. The exact location and date of chicken domestication are still debated, but some evidence suggests that it may have occurred in multiple places and times, such as the Indus Valley, China, and Southeast Asia, between 7,000 and 10,000 years ago. From there, chickens were introduced and dispersed to other regions, such as the Middle East, Europe, Africa, and the

Americas, through trade, migration, and colonization.

- **Explore the variations:** The final step to trace the origins and development of chickens is to explore the variations, or the differences and similarities among chicken breeds and populations. The variations of chickens are the result of natural and artificial selection, as well as genetic drift and gene flow, that shaped the chicken's appearance, behavior, and performance. There are hundreds of chicken breeds and varieties in the world, each with its own characteristics, such as size, shape, color, comb, feather, egg, and meat.

Understanding the History and Evolution of Chickens

The history and evolution of chickens are the stories and changes of chickens over time. They involve the events and trends that affected the role and status of chickens in human society and culture. Some of the main periods and aspects of the history and evolution of chickens are:

- *Ancient times:* In ancient times, chickens were mainly used for religious, ceremonial, and entertainment purposes, such as sacrifices, divination, and cockfighting. Chickens were also valued for their symbolic and spiritual meanings, such as courage, fertility, and wisdom. Chickens were rarely eaten, as they were considered sacred or scarce, and eggs were a luxury item.

- *Medieval times:* In medieval times, chickens became more common and accessible, as they were bred and kept by peasants, monks, and nobles. Chickens were also used for practical and economic purposes, such as food, feathers, and manure. Chickens were still involved in some religious and cultural practices, such as Easter, Christmas, and carnival.

- *Modern times:* In modern times, chickens became more industrialized and commercialized, as they were mass-produced and consumed by the growing human population.

Chickens were also improved and diversified, as they were selectively bred and genetically modified for various traits, such as egg production, meat quality, and disease resistance. Chickens were still part of some social and ethical issues, such as animal welfare, environmental impact, and food security.

Appreciating the Genetic Diversity and Cultural Significance of Chickens

The genetic diversity and cultural significance of chickens are the features and values of chickens that make them unique and important. They involve the biological and social factors that influence the variation and appreciation of chickens. There are Here ways to appreciate the genetic diversity and cultural significance of chickens which includes:

- *Compare the genetics:* One way to appreciate the genetic diversity of chickens is to compare their genetics, or the molecular and hereditary basis of their traits and characteristics. You can use tools such as DNA testing, phylogenetic analysis, or genome sequencing to examine and compare the genetic makeup and relationships of different chicken breeds and populations. You can also use tools such as phenotypic evaluation, breed standards, or trait measurements to observe and compare the physical and behavioral expressions of different chicken genes.

- *Celebrate the culture:* Another way to appreciate the cultural significance of chickens is to celebrate their culture, or the human and social context of their meaning and importance. You can use sources such as history, literature, art, or folklore to learn and appreciate the role and influence of chickens in different civilizations, religions, and traditions. You can also use sources such as cuisine, fashion, or hobby to enjoy and appreciate the taste and beauty of chickens in different regions, styles, and preferences.

Conclusion

You have reached the end of this guide on how to keep chickens in your backyard or farm. Congratulations! You have learned a lot of information and skills on how to care for your chickens, how to protect them from predators and diseases, how to collect and store their eggs, how to interact and bond with them, and how to understand their behavior and intelligence. You have also discovered their history and evolution, their genetic diversity and cultural significance, and their anatomy and physiology. You have become a confident and competent chicken keeper!

But your chicken-keeping journey does not end here. There is always more to learn and improve, more to share and celebrate, and more to express and appreciate. Here are some tips on how to review and reflect on your chicken-keeping journey, how to continuously improve and develop your skills, how to share and celebrate your experiences, and how to express gratitude and love for your chickens.

Review and Reflection on Your Chicken-Keeping Journey

Review and reflection are the processes of looking back and thinking about your chicken-keeping journey. They help you evaluate your progress, achievements, and challenges, and also identify your strengths, weaknesses, and opportunities. Here are some steps to follow when reviewing and reflecting on your chicken-keeping journey:

- *Recall:* Recall your chicken-keeping journey from the beginning to the end. Remember the reasons why you decided to keep chickens, the goals you set for yourself, the actions you took, the results you obtained, and the feedback you received.

- *Analyze:* Analyze your chicken-keeping journey in terms of what worked well and what did not work well. Consider the factors that contributed to your success and failure, such as your knowledge, skills, attitude, resources, and environment. Also consider the impact of your chicken-keeping journey on yourself, your chickens, and others, such as your satisfaction, happiness, learning, and growth.

- *Learn:* Learn from your chicken-keeping journey by drawing lessons, insights, and recommendations. Think about what you have learned about yourself, your chickens, and your chicken-keeping practice. Think about what you can do better, differently, or more in the future. Think about what you can share, teach, or advise others who are interested in chicken-keeping.

Continuous Improvement and Skill Development

Continuous improvement and skill development are the processes of enhancing and expanding your chicken-keeping practice. They help you maintain your quality, efficiency, and effectiveness, and also explore your potential, creativity, and innovation. The following are some ways to achieve continuous improvement and skill development:

- *Update:* Update your chicken-keeping knowledge and skills by staying informed and educated. You can use sources such as books, magazines, websites, blogs, podcasts, or videos to learn about the latest trends, research, and innovations in chicken-keeping. You can also use sources such as courses, workshops, seminars, or webinars to learn new techniques, methods, or approaches in chicken-keeping.

- *Experiment:* Experiment with your chicken-keeping practice by trying new things and taking risks. You can use tools such as puzzles, tricks, or games to challenge and stimulate your chickens' intelligence and learning abilities. You can also use tools such as toys, treats, or rewards to enrich and diversify your chickens' lives and minds.

- *Network:* Network with other chicken-keepers by connecting and collaborating. You can use platforms such as forums, groups, clubs, or associations to meet and interact with other chicken-keepers. You can also use platforms such as events, shows, fairs, or competitions to participate and compete with other chicken-keepers.

Sharing and Celebrating Your Experiences
Sharing and celebrating are the processes of expressing and enjoying your chicken-keeping experiences. They help you communicate and connect with others, and also appreciate and acknowledge your efforts and achievements.

You can keep, share and celebrate your experiences through:

- *Document:* Document your chicken-keeping experiences by recording and organizing them. You can use media such as photos, videos, or audio to capture and preserve your chicken-keeping moments. You can also use media such as journals, diaries, or calendars to document and track your chicken-keeping activities.

- *Showcase:* Showcase your chicken-keeping experiences by displaying and presenting them. You can use platforms such as social media, blogs, or websites to share and promote your chicken-keeping stories, tips, or advice. You can also use platforms such as galleries, exhibitions, or portfolios to showcase and demonstrate your chicken-keeping skills, products, or services.

- *Celebrate:* Celebrate your chicken-keeping experiences by rewarding and congratulating yourself and your chickens. You can use methods such as treats, gifts, or parties to treat and spoil yourself and your chickens.

You can also use methods such as certificates, medals, or trophies to recognize and honor yourself and your chickens.

Expressing Gratitude and Love for Your Chickens

Expressing gratitude and love are the processes of showing and feeling appreciation and affection for your chickens. They help you build and strengthen your relationship and bond with your chickens, and also foster and nurture your happiness and well-being. Here are some ways to express gratitude and love for your chickens:

- *Thank:* Thank your chickens for being part of your life and for giving you joy, eggs, and companionship. You can use words, gestures, or signals to communicate and convey your gratitude and appreciation to your chickens. You can also use actions, behaviors, or routines to demonstrate and prove your gratitude and appreciation to your chickens.

- *Care:* Care for your chickens by providing them with the best possible living conditions and care. You can use resources such as food, water, shelter, bedding, and health care

to ensure and support your chickens'
physical and mental health and comfort.

You can also use resources such as toys, treats, and enrichment to enhance and stimulate your chickens' happiness and satisfaction.

- **Cuddle:** Cuddle with your chickens by holding and hugging them. You can use your hands, arms, and lap to embrace and comfort your chickens. You can also use your fingers, nails, and lips to pet and kiss your chickens. You can also feel their warmth, softness, and heartbeat, and share your comfort and love with your chickens.

CHICKEN FEEDING JOURNAL

THIS BOOK BELONGS TO:

Chicken Feeding Journal

FOR THE WEEK

MONDAY

TUESDAY

WEDNESDAY

THURSDAY

FRIDAY

SATURDAY

Chicken Feeding Journal

FOR THE WEEK

MONDAY	TUESDAY

WEDNESDAY	THURSDAY

FRIDAY	SATURDAY

Chicken Feeding Journal

MONDAY

TUESDAY

WEDNESDAY

THURSDAY

FRIDAY

SATURDAY

Chicken Feeding Journal

FOR THE WEEK

MONDAY	TUESDAY
WEDNESDAY	THURSDAY
FRIDAY	SATURDAY

Chicken Feeding Journal

FOR THE WEEK

MONDAY	TUESDAY

WEDNESDAY	THURSDAY

FRIDAY	SATURDAY

Chicken Feeding Journal

FOR THE WEEK

MONDAY	TUESDAY
WEDNESDAY	THURSDAY
FRIDAY	SATURDAY

Chicken Feeding Journal

FOR THE WEEK

MONDAY

TUESDAY

WEDNESDAY

THURSDAY

FRIDAY

SATURDAY

Chicken Feeding Journal

FOR THE WEEK

MONDAY	TUESDAY
WEDNESDAY	THURSDAY
FRIDAY	SATURDAY

Chicken Feeding Journal

FOR THE WEEK

MONDAY	TUESDAY

WEDNESDAY	THURSDAY

FRIDAY	SATURDAY

Chicken Feeding Journal

FOR THE WEEK

MONDAY	TUESDAY
WEDNESDAY	THURSDAY
FRIDAY	SATURDAY

Chicken Feeding Journal

FOR THE WEEK

MONDAY	TUESDAY
WEDNESDAY	THURSDAY
FRIDAY	SATURDAY

Chicken Feeding Journal

FOR THE WEEK

MONDAY	TUESDAY
WEDNESDAY	THURSDAY
FRIDAY	SATURDAY

140

www.ingramcontent.com/pod-product-compliance
Lightning Source LLC
Chambersburg PA
CBHW070033300526
45794CB00001B/475